W9-DAO-009

THE INCREDIBLE WORLD OF PLANTS

PLANTS OF THE FOREST

CHELSEA HOUSE PUBLISHERS
New York • Philadelphia

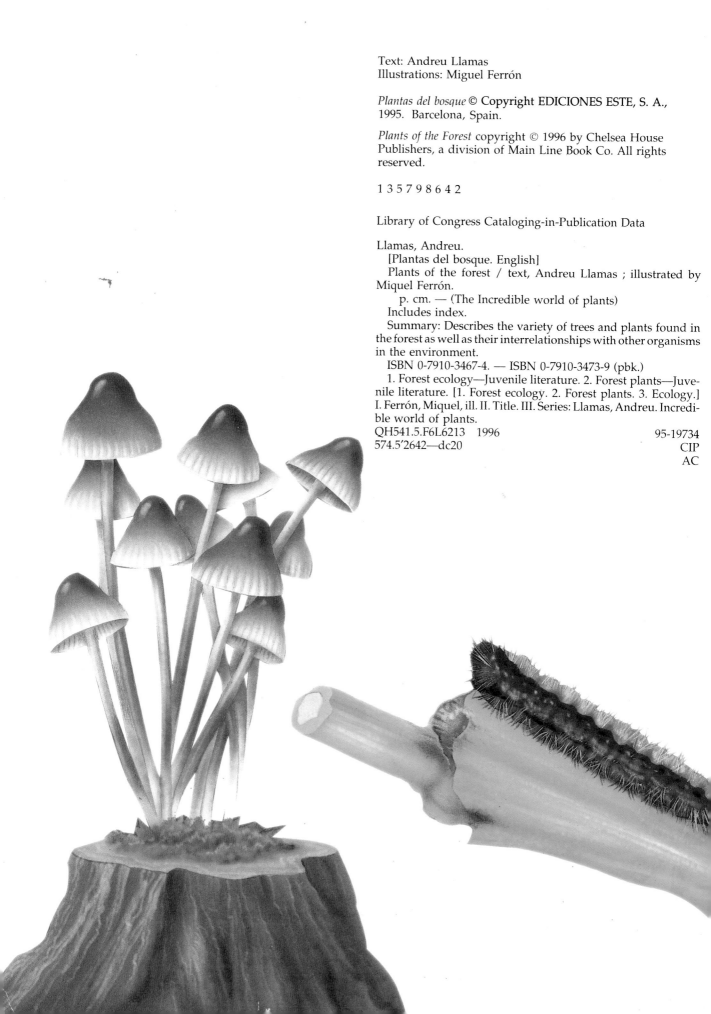

Text: Andreu Llamas
Illustrations: Miguel Ferrón

Plantas del bosque © Copyright EDICIONES ESTE, S. A.,
1995. Barcelona, Spain.

Plants of the Forest copyright © 1996 by Chelsea House
Publishers, a division of Main Line Book Co. All rights
reserved.

1 3 5 7 9 8 6 4 2

Library of Congress Cataloging-in-Publication Data

Llamas, Andreu.
  [Plantas del bosque. English]
  Plants of the forest / text, Andreu Llamas ; illustrated by
Miquel Ferrón.
    p. cm. — (The Incredible world of plants)
  Includes index.
  Summary: Describes the variety of trees and plants found in
the forest as well as their interrelationships with other organisms
in the environment.
  ISBN 0-7910-3467-4. — ISBN 0-7910-3473-9 (pbk.)
  1. Forest ecology—Juvenile literature. 2. Forest plants—Juve-
nile literature. [1. Forest ecology. 2. Forest plants. 3. Ecology.]
I. Ferrón, Miquel, ill. II. Title. III. Series: Llamas, Andreu. Incredible world of plants.
QH541.5.F6L6213   1996   95-19734
574.5′2642—dc20   CIP
                                                              AC

CONTENTS

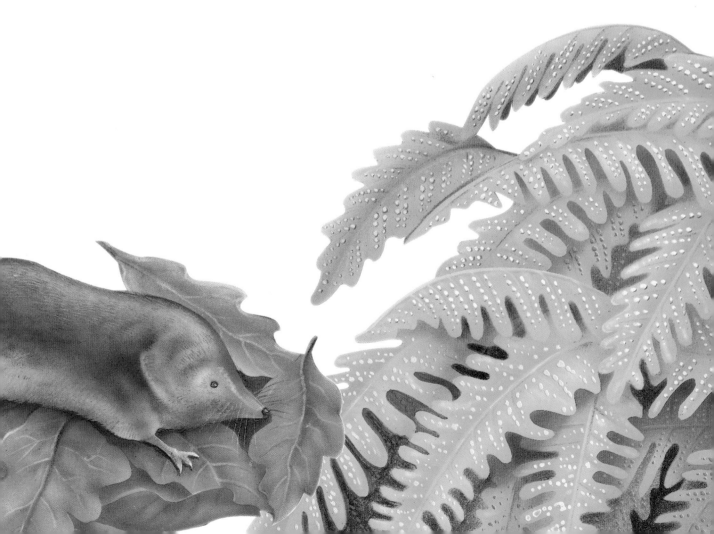

THE TREES IN THE FORESTS

Forests are very important on earth—they cover 40 percent of the land, excluding deserts and the North and South poles.

Forests are land ecosystems dominated by trees. Every tree produces thousands of seeds, but each one takes a long time to grow into a tree and form a forest with others. The trees that grow also have to fight for space, water, and light, so the weaker trees die.

Forests are made up of "floors" or *strata*. The light that reaches the treetops manages to filter through the leaves and reach the vegetation below. This is why vegetation forms different strata.

However, not all tress have the same strata below them. Beech trees, for instance, have a very thick leaf covering and let through little light, so few plants are able to grow below them. Thin needles of pine trees, on the other hand, allow a lot of light to get through and many bushes grow below these trees.

Each plant tries to catch as much sunlight as possible, but light is scarce in the forests and all plants want it, so there is a constant fight for the best-lit places.

Each tree is a little universe, and an enormous number of animal and vegetable species can be found there. On the bark of a single oak there might be as many as 30 different species of lichen.

There are also many fungi and *invertebrates* that live on dead and rotting wood; others prefer to live in the cloak of leaves that fall every autumn.

(1) The small world of the tree
A forest is like a miniature world, and each tree is also a small world, where many animals and plants can be found.

(2) Shield bugs
The shield bugs have enormous jaws that look like deer horns. They feed by sucking on the fluid of plants or animals.

(3) Woodpeckers
Very often in the forest we hear the hard knocking of woodpeckers on a tree when they are building a nest or looking for hidden larvae. This noise can be heard up to a half mile away.

(4) An animal drill
The ichneumon Rhyssa persuasoria can lay its egg on an insect larva even if it is hidden several inches inside a trunk. It first detects the larva through the wood with its antennae, and then it uses the back of its body like a drill to make a hole and reach the larva.

(5) Borers or hatchers
The female makes long passageways in the bark of sick trees. It builds small niches and leaves an egg in each one. Later, the larvae that hatch from the eggs gnaw the wood to feed, and when they are adults they make a hole to get out.

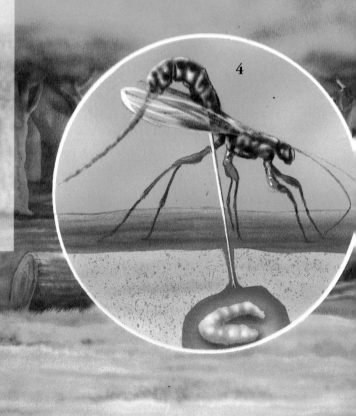

4

SPRING-SUMMER DECIDUOUS FORESTS

When spring comes, the sun warms the ground for a long period of time, so plants on the ground hurry to flower before the branches of the trees once again become covered with leaves.

Deciduous forests are made up of trees that simultaneously lose all their leaves each autumn and grow them again in the spring. These types of forests abound in mild climates, and the most typical species of trees that are found here are beeches, oaks, chestnuts, and birches.

When the spring sun shines and the trees do not yet have leaves, the forest is full of light. This is the time for plants on the ground to get the light they need before the leaves cover them in half-light once again.

Later, the trees quickly cover with leaves. The leafy cover is so dense that sunbeams cannot get through, so the only plants that survive in the *undergrowth* are those that need little light.

Many animals also wake when spring arrives and the forest is full of activity. Leaves appear on the branches and with them caterpillars ready to eat them. Some trees try to defend themselves by accumulating chemical repellents in their leaves, but this does not always help.

(1) Life awakens
In spring, life fills the forest. Animals and plants make use of the good weather to feed and reproduce. This great activity lasts until the end of summer.

(2) Weevils
This weevil has a long, thin beak, which it uses to make a small hole in acorns in the summer. This is how it puts its eggs inside the beak.

(3) The tit bird
The tit catches several caterpillars to feed its many children. You can see it catching a caterpillar that was eating an ilex leaf.

(4) Camouflaged
The bucephalus caterpillar feeds at the end of the summer. When it is an adult, it turns into a butterfly that looks like rotting wood, so it throws off its predators.

(5) Butterflies
In the summer, butterflies fly all around. Here you can see a forest nymph (top) and a thyme butterfly (bottom).

(6) Moths
Most moths rest during the day, camouflaged. The beech warrior moth is active from the end of spring to the beginning of summer.

5

6

DECIDUOUS FORESTS IN AUTUMN

When autumn comes, as the days shorten and there are fewer hours of light, forests transform and all the trees begin to lose their leaves.

In no other place are the changing seasons so visible as in deciduous forests, because throughout the year they change color: soft green in spring, luxuriant in summer, golden and blazing in autumn, and finally, bare in winter.

When autumn comes, the forest fruit and seeds mature. Remember that trees want their seeds to be spread as far as possible, so they have invented many systems for travel. For instance, some seeds have extensions that, like wings, help them to be taken as far as possible in the wind. Other seeds have tiny hooks to catch on to the skin of mammals, and they travel as stowaways!

Fruit also attract animals with their brilliant colors. When an animal eats the fruit, its hard seeds pass through its body undigested and fall a long way from their tree. Animals take advantage of the enormous abundance of food in autumn; they accumulate part of this food in their bodies as fat layers, which will be a good energy stock to help them through the hard winter. Acorns are one of the most important foods of forest mammals and birds. Just think that one oak can produce more than 50,000 acorns in one year.

Halfway through autumn, the leaves begin to fall and gather in a layer called *humus*. The result is a protective layer which gives food and shelter to many small animals.

(1) Autumn arrives
Autumn transforms the colors of deciduous forests. The surroundings that were once green are gradually invaded by reds and yellows.

(2) Changes in the leaves
Here you can see the changes to leaves during the autumn. In the end, the dead leaf falls off the tree. This change occurs because the green chlorophyll in leaves disappears when there are fewer hours of light.

(3) Warming bodies
The procustes beetle remains active even when it is cold, but prefers to hunt in the sunlight. The plate on its back is used to catch and store the sun's rays. Its whole body is black to hold the warmth of the sun.

(4) Tasty fruit
When autumn comes, bushes and climbing plants show their fruit, such as these tasty blackberries.

(5) "Dead leaves"
The preussia, also known as the leaf grasshopper, can camouflage itself perfectly. As you can see, its whole body looks just like a dead leaf.

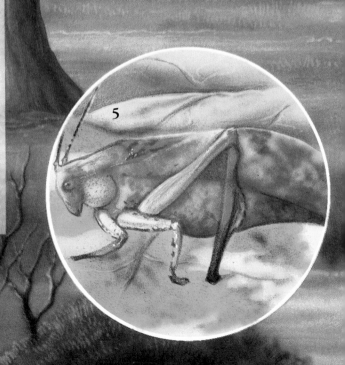

5

DECIDUOUS FORESTS IN WINTER

Winter is a time of rest for the forest, which is quiet and lifeless.

Plants stop growing, flowering, and giving fruit during the winter. You have already seen that temperate forests are deciduous and that in the winter they lose their leaves to "rest." However, they are not the only life forms to take a break—many animals in the forest *hibernate* in hidden places to escape from the cold, spending the winter asleep.

During the winter, the bare trees stop growing completely because they have no leaves for *photosynthesis*. This is why we can see the growth of trees in their rings (the number of rings in a stump is equivalent to the age of the tree).

Snowfalls may be very intense, but snow is beneficial for forests in winter because the snow that falls on the ground acts as a blanket and protects the vegetation and creatures from the cold. However, sometimes an excessive snowfall may weigh too much for the branches of some trees and can break them. This is why trees that grow in areas of heavy snowfall have special branches that do not accumulate snow.

Despite everything that happens in the winter, life does not stop for the inhabitants of the forest, and if you look carefully at the snow on the ground, you will see the footprints of animals that are still active, such as squirrels. Many of these survive thanks to the stock of food they hide during the autumn.

(1) The forest sleeps
In winter, life seems to come to a standstill in the forest. The days are short and gray, and the cold can be very intense.

(2) Butterflies in winter
We may be surprised to see adult butterflies when out walking in the forest during the winter, but this is when butterflies emerge from their underground chrysalises.

(3) Sleeping in winter
During the winter, many animals, such as this small dormouse, prefer to hibernate.

(4) A good refuge
Many eggs, larvae, and caterpillars spend the winter in protected places, such as inside tree trunks.

(5) A warm habitation
The earthworm spends the winter rolled up on the ground.

5

1

2

3

4

11

PINE FORESTS

Conifers grow all around the world, but especially in cold regions, because they are best adapted to this kind of climate.

Conifers are very strong and able to survive during drought, low temperatures, snow, and wind. Their conical shape also helps them to stand the strength of the wind and the weight of the snow. The best known species of conifer are pines, firs, spruces, larches, and birches.

There are about 100 different species of pines, and they are special because of their hard, narrow leaves covered in wax to prevent excessive sweating and to retain water.

Almost all species of evergreen produce pine cones instead of flowers. Most species grow to be over 66 feet (20 meters) high, some are more than 164 feet (50 meters), and the giant species, almost exclusive to the west coast of the United States, reach 328 feet (100 meters) high.

The *processionary caterpillars* of the pine are not the only insects that feed on the acicular leaves. There is also a group of sawing flies that eat this kind of leaves. There are some insects that enter the wood while other mites prefer to suck the resinous sap.

The pine forest floor is covered with a layer of leaves, twigs, and pine cones, which decompose slowly, because the decomposing bacteria find it difficult to work with in the cold of these forests.

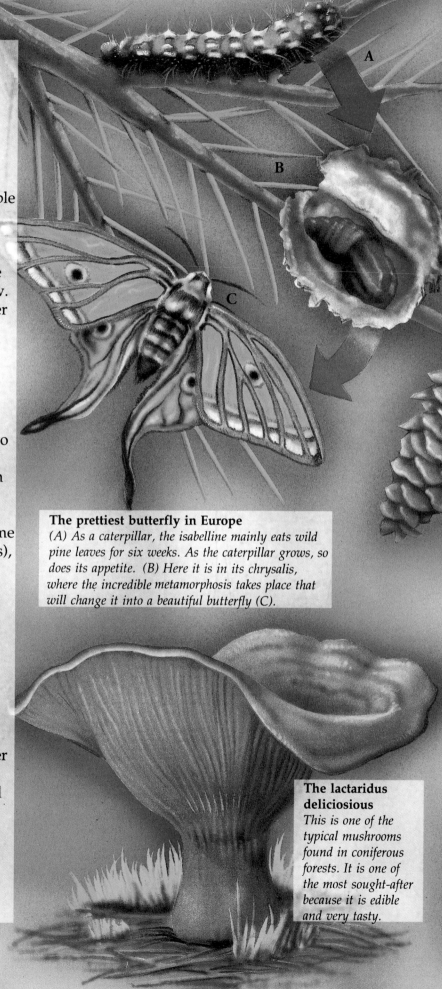

The prettiest butterfly in Europe
(A) As a caterpillar, the isabelline mainly eats wild pine leaves for six weeks. As the caterpillar grows, so does its appetite. (B) Here it is in its chrysalis, where the incredible metamorphosis takes place that will change it into a beautiful butterfly (C).

The lactaridus deliciosious
This is one of the typical mushrooms found in coniferous forests. It is one of the most sought-after because it is edible and very tasty.

Enormous pine cones
The biggest cones are produced by the Lambert pine tree in California. They may be more than 20 inches long and weigh over 1 pound!

Parasitic mushrooms
There are many fungi in coniferous forests, and most live as parasites on the trees. Their shape and size can vary enormously: (A) Gomphydius, (B) Boletus reticulatus, and (C) Mycena seynii.

Different conifers
There are many different species of conifer, of which these are three. You will see there are great differences between them: (A) the fir, (B) the pine, and (C) the savin.

13

THE FOREST FLOOR

Did you know that more than a ton of leaves fall on every acre of forest floor in autumn?

The forest floor is a small universe made up of dead leaves, twigs, moss, toadstools, acorns, and flowers. In this complicated vegetable cloak live small vertebrates such as the shrew and many invertebrates such as red ants, spiders, and centipedes.

All vegetable material that falls on the forest floor decomposes and changes into a dark, rich earth called humus. The wood and leaves are changed into humus by millions of tiny inhabitants on the floor. In 4 ounces (100 grams) of soil there may be more than 100,000 bacteria! The wood is attacked by fungi and the leaves are chewed by, among others, the enormous armies of acari and wood lice.

Also in the undergrowth earthworms mix soil from different levels. Finally, microbes and fungi carry out the final stages of decomposition of the vegetable cloak.

By producing humus, the forest enriches the soil constantly. It is quite a slow process—the leaf layer takes nine months to decompose completely to make humus. This is how the nutrients used by the trees, such as phosphorus and nitrogen, are returned to the soil. Remember that life does not stop in the forest during the winter. Snow acts as an insulator, and the creatures below it continue their activity.

(1) Leaves everywhere
This is what the floor looks like in a deciduous forest. The fallen leaves cover almost the entire surface of the forest during their transformation into humus.

(2) Growing on fallen trunks
Many fungi, such as these hypholoma (2a) and tinder fungi (2b), grow on the wood from fallen trunks.

(3) Enormous larvae
Stag beetle larvae spend three years feeding inside the rotting wood of a fallen tree, until they are 5 inches (12 centimeters) long.

(4) Wood lice
Wood lice prefer to live in damp and shady surroundings. If you want to find some, you should look under rocks.

(5) Nests underground
Some birds can have their nests underground, like the coal tit, which has its nest inside an abandoned underground mouse hole.

(6) False scorpions
Although these animals look like scorpions they are really inoffensive pseudoscorpions.

(7) The tiny shrew
The shrew is the smallest mammal, weighing less than .07 ounces! It moves through the foliage, turning the leaves over with its nose looking for small invertebrates. Every night it hunts for spiders and insects because each day it needs to consume more food than its own bodyweight.

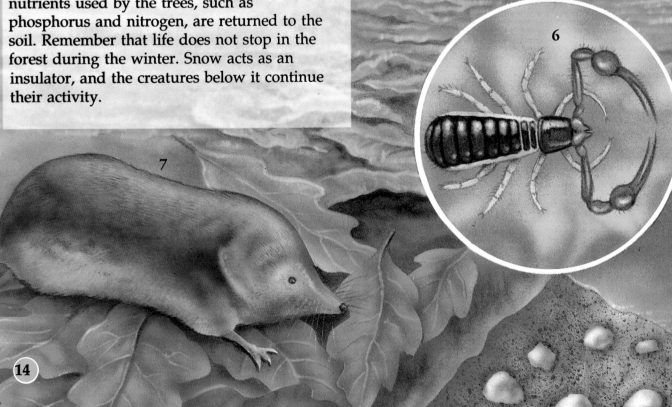

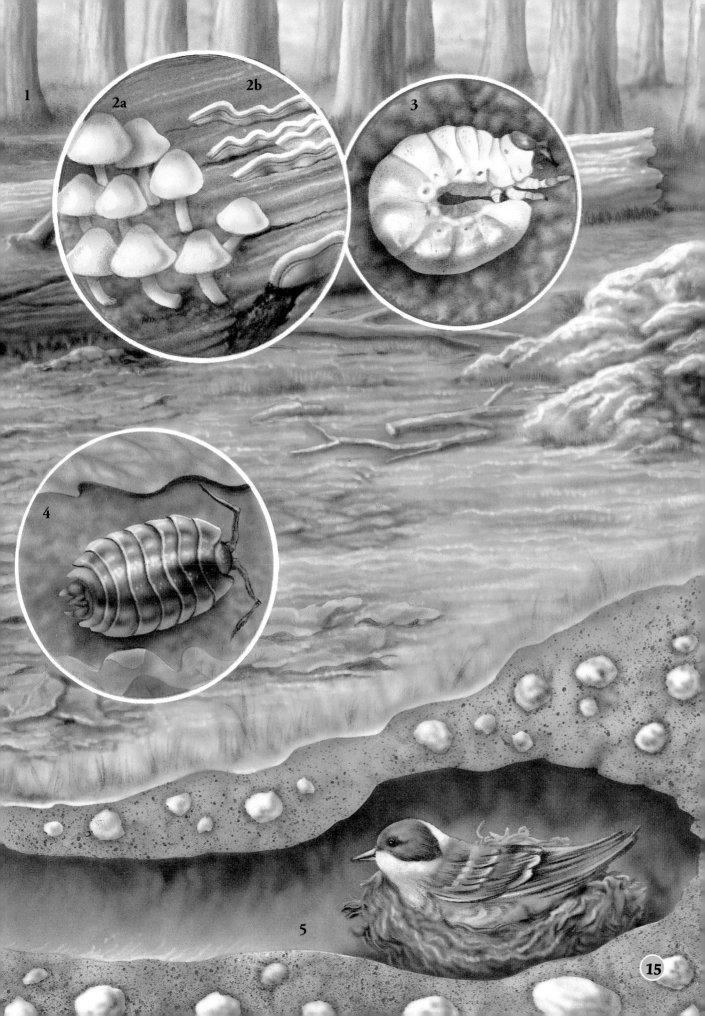

1

2a 2b

3

4

5

15

THE PINE PROCESSIONARY

You may have seen the great white bags that the processionary caterpillars build on pines.

A month after laying their eggs, at the end of summer, caterpillars hatch and build their sacks on tree branches. There they hide during the day and come out at dusk to eat the pine needles of the trees. They can cause tremendous damage in forests.

The most striking thing about these insects are their long processions. At the beginning of spring, the caterpillars come down tree trunks in an orderly procession and look for a place to bury themselves so they can change into adults.

There is always a female at the head of the procession. If she is eliminated, two things can happen: if the following individual is female, the column will continue calmly with a new guide at the front; if the leading individual is male, it begins to go from one side to another trying to make contact and the whole column is thrown off and piles up until another female takes the initiative!

Although caterpillars seem inoffensive, they have small lumps on their backs called mirrors, covered in small folds of the *integument*. When they feel threatened, caterpillars uncover these mirrors and free small, irritating hairs that cause soreness and unpleasant allergic reactions if they touch the skin of an animal.

Caterpillars feed untiringly on the hard, narrow leaves called acicula. They can often leave a tree without almost any leaves—they can even kill it.

Winter processionary adults appear in the summer, lay their eggs quickly, and die. This is what the processionary clutch looks like.

As refuge, the caterpillars use characteristic sacks that hang from the end of pine branches.

Although many people have seen processionary caterpillars in their columns, very few are able to say what they look like as adults. Here you can see an adult couple.

This calosoma beetle is capable of eating caterpillars. It opens the caterpillar with its jaws and puts a substance on them to soften the victim's flesh. Then it only has to drink the caterpillar soup without having to swallow any hairs.

Processionary caterpillars always live together, eat together, and move together from one tree to another in long columns of up to 33 feet. Caterpillar processions take place about three months after they hatch.

THE TAIGA

Conifer forests cover more than 8,100 miles. Life on the planet depends on this great forest mass, which is the second largest manufacturer of oxygen after oceans.

The *taiga* covers areas around 60 degrees latitude, where the tundra ends. Conditions are very hard here, as there is very little water available for most of the year and it is extremely cold. Actually, the word taiga means "lake forest" and it is true that the countryside consists of thousands and thousands of lakes, but most of them are frozen for more than half the year. It is difficult to walk in these forests because there are constantly small lakes or frosty areas where peat bogs abound.

However, the cold is an even greater problem, because in the taiga the temperature often falls to 58 degrees below zero Fahrenheit. The soil of the taiga is very poor, and for all these reasons the taiga is a unit of almost interminable forests made up of only one family of trees that have adapted to the extreme conditions—the conifers—represented especially by pines, firs, spruces, and larches (which can lose all their leaves). These trees are very strong and can stop growing during the long bad season. In fact, they only grow in the summer.

The undergrowth is made up of bushes, such as heather and bilberry bushes, but usually it is fairly poor due to the lack of light and the slowness of rotting processes. There are many species of animals, especially birds, that visit these regions during the five months when temperatures exceed the freezing point (although it never rises above 60 degrees).

(1) Monotonous scenery
Although it is immense, the taiga can sometimes seem monotonous because it is only made up of a few species of different trees. The soil in the taiga is very poor, and conifers survive because they make few demands for nutrients.

(2) A very strange beak
The crossbill is perfectly adapted to life on the taiga. Its beak is bent and crossed and makes an ideal tool for extracting the contents of the very hard pine cones.

(3) The blue climber
Climbers can stroll peacefully up and down the vertical surfaces of the trees thanks to the claws on their long fingers. In this way they can look in all the nooks and crannies for hidden insects.

(4) What a memory!
During the long and hard winters, the nutcracker feeds from the stocks of food it has hidden during summer and autumn. It can find where it has hidden food after several months, even if the food is covered by 5 feet of snow!

(5) Butterflies with cyanide
This species of butterfly is very slow, it has striking colors, and its sexual act lasts a whole day. So, how does it defend itself from its enemies? Its body contains a very powerful poison—cyanide.

(6) The apollo butterfly
These butterflies can live over 6,500 feet (2,000 meters) high and are legally protected in some countries.

6

MEDITERRANEAN FORESTS

The trees and bushes of Mediterranean forests are not usually more than 66 feet (20 meters) high and they keep their leaves all year around despite the droughts.

Mediterranean forests are in regions with little rain, mild winters, and hot summers. It is the typical scenery around the Mediterranean Sea but also appears on the California coast, in southern Australia, and in South Africa. To survive during the long droughts of summer, trees in Mediterranean forests have special leaves that sweat very little. But the price of their resistance is that they grow slowly.

The typical Mediterranean trees are the ilex, cork oak, olive, cypress, and eucalyptus trees. The forests consist of the three typical layers: sylvian, shrub, and herbaceous.

Generally, the leaves of Mediterranean trees let through a lot of light to the lower layers, so the shrub layer forms a thick undergrowth.

There are also many climbers that grow among the bushes and trees and join their branches to make the forest impenetrable. It is difficult to walk in these forests, but the animals that live there have adapted to moving among the tangle of bushes.

The herbaceous layer is thin due to the small amount of light that reaches the ground. There, the surroundings are damp, which facilitates growth of moss, ferns, and some herbaceous plants such as violets.

(1) The ilex grove
Ilexes are the most common trees in Mediterranean forests. They are large trees up to 82 feet (25 meters) in height.

(2) The ilex fruit
The acorn, generally oval-shaped and pointed, is the fruit of the ilex and the oak, among other trees. Its flavor is sweet or bitter.

(3) Cork
The cork oak does not grow more than 66 feet (20 meters) but can reach 3 feet (1 meter) in diameter. The cork oak has spongy bark, which provides us with cork.

(4) The flower covering
Have you seen the perfect camouflage of this hunter? It must be terrible for its victims when they discover that it is not a flower but a terrible predator with spikes on its legs to catch its prey better.

(5) The "cigar" weevil
The female prepares a case in which to keep her eggs by rolling leaves like a big cigar. First she chooses a flexible leaf and then cuts it, sticking some parts to others with a substance she secretes. The larvae develop by eating the walls of their case when they leave the eggs.

(6) Beneficial ants
Ants are one of the most beneficial animals in the forest, as they exterminate a great number of insects. In one day, an army of ants can capture 50,000 larvae of dangerous insects.

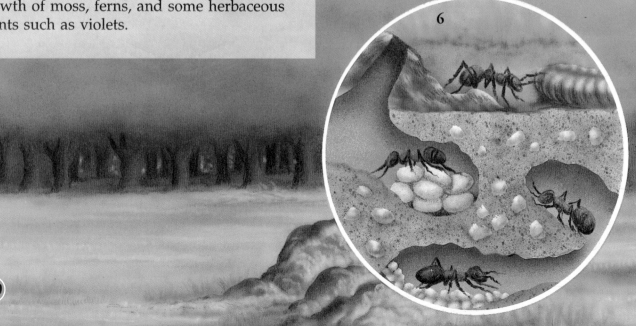

6

HOW DO FOREST TREES GROW?

Through its life, every tree produces thousands—even millions of seeds, but very few become adult trees.

Forests need many years to form, despite the fact that trees make a tremendous effort to reproduce. In one year, each oak tree may produce more than 50,000 acorns. If it lives long enough, an oak tree may have produced millions of acorns, though very few of them become trees.

When the seeds have not yet fallen off the branches, they are attacked by animals such as squirrels, weevils, crossbills, caterpillars, and wood pigeons. Thousands of pine cones, nuts, and acorns, are attacked before they can even begin their journey.

The seeds that manage to get to the ground face new difficulties: many of them fall in places where they cannot grow (too much light or shade, few nutrients, or competition), and others are eaten by deer, boars, and rodents that run around the forest floor.

Inside the seeds there are nutrients that provide them with the energy they need to germinate and grow, but in some cases, such as with oak and beech seeds, the stock is sufficient for them to survive the whole winter.

To grow well, the seeds need a little shade, but not too much (as they would be attacked by fungi).

Burier beetles
Burier beetles eat the dead animals they find on the forest floor and lay their eggs next to the corpses. To protect it from other insects, they bury the corpse by digging a hole under it.

1. From the end of spring to the middle of autumn, acorns develop on oak branches.
2. The acorn falls to the floor and is covered by the leaves that fall off the tree.
3. The main root comes out of the acorn directly downward, and the shoot grows toward the light.
4. If it manages to survive the first year it already has some leaves. This stage is very dangerous because many are devoured by browsers.
5. These first years are very dangerous since the peril of being eaten will not pass until they are 10 or 15 years old.

1

2

3

It is not a leaf
When settled, the kalima butterfly folds its wings and looks like a leaf.

The crane or reaper
Thanks to its long legs, the crane or reaper moves faster than other animals of its size.

5

FUNGI

During the damp, warm days of autumn, the forest floor becomes covered with fungi. If you walk through the forest in autumn you will see an abundance of mushrooms.

There are many different environments in the forests where plants can grow without flowers, both on the floor and on the trunks of living and dead trees. Trees give fungi protection from sunlight, wind, and rain. Mushrooms are very strong decomposers—wherever the wood is mossy they begin to work. Fungi have no chlorophyll and so they cannot use the sun's energy to produce food by photosynthesis. So, many fungi in forests live as parasites on the branches and trunks of living trees, while others only live on their remains, on dead stumps and trunks. The known mushrooms are the visible fruit of fungi. The main part of a fungus is the mycelium, which is made up of a type of net with many tiny, microscopic threads called hyphae.

Some fungi that live on the ground may come to a very special "agreement" with tree roots. These fungi infect the tree roots and form an association that is beneficial both for the tree and for the fungi. Thanks to the fungi, the tree improves its absorption of nutrients, and in exchange, the tree gives the fungi the carbohydrates they need.

Growing in small groups
The mycete is a nonedible fungus that grows in small groups on stumps and dead trunks.

The polypary
The sulphur polypary has a striking yellow color and lives on the trunks of living trees. Its flesh is edible, though not very tasty.

A giant snail
In Asia and Africa there is a snail whose shell can measure up to 8 inches by 4 inches. It is very hungry and can cause a lot of damage in plantations, but it is so big that it makes a great feast for some animal.

Don't touch!
This is the Amanita phalloides. It is very toxic and is responsible for most fatal poisonings caused in Europe through eating fungi.

"Ox liver"
This fungus can be found at the base of chestnut trees or oaks. It gets its name from its pink color and the fact it is quite soft and juicy. It also secretes a red liquid when it is knocked or cut.

Pretty but toxic
The Amanita muscaria has a strange bright red hat, but watch out— it is toxic and has been used at certain times and places to cause hallucinations.

GILLS

Sometimes we can see strange formations that grow on parts of a tree. These are gills and are where the larvae of small creatures such as insects and fungi live.

Gills are odd structures that appear on some plants. It is an abnormal growth of the plant, in response to stimuli from other organisms such as insects, bugs, fungi, or bacteria. Gills are a tree defense mechanism against intruders, normally in the shape of the larvae of a small wasp.

But why do insects cause the appearance of a gill? The fact is that the larvae use this abnormal growth to feed within the protection of the gill, which gives them a sure source of food so the larvae grow surrounded by the nutritious tissue of the guest plant and are also protected from their natural enemies and the climate outside.

Gills caused by the same species of insect may appear on different parts of the same tree and even on different species of trees.

The idea of gills has been so successful that there are more than 40 species that can cause gills to appear on oaks. In this case, most of them are caused by the tiny wasps of the cynapid family of gills.

Many species have developed systems to take advantage of the same benefits. This is why some gills even house a complicated community of parasites, predators and lodger wasps. It is like a housing block full of neighbors.

Marble-shaped gill
This is one of the best-known gills and is produced by the wasp Andricus collari. When the wasps leave the gill, they leave characteristic round holes.

Artichoke-shaped gill
Despite its appearance, it only contains one larva of the gill wasp Andricus fecundator.

Knopper gill
This strange gill is produced by the wasp Andricus

Cotton gill
It is like a small ball of cotton 5 inches in diameter and may be made up of many individual gills together (up to 20).

Worm or "oak apple"
As you can see, this gill really looks like an apple, but it has not formed on an apple tree but on an oak.

Cherry-shaped gills
By looking at this picture, you can see that these gills look curiously like cherries.

FOREST GIANTS: REDWOODS

There are some trees that possess all the records for age and size in the vegetable world: the redwoods.

The giant redwoods grow on mountainous slopes where there is moisture from the ocean mists. If there are no excessively strong winds in this area, there are no obstacles to growth, and redwoods can reach impressive heights. Together with eucalyptus trees, redwoods are the tallest trees in the world at over 400 feet (120 meters) tall. Their trunks can also be more than 40 feet (12 meters) wide—that is three cars lined up front to back.

Redwoods are also some of the oldest inhabitants of the planet; the age of some of these trees has been calculated at more than 3,000 years.

In ancient times, before the ice ages, the redwoods formed great forests all over the Northern Hemisphere, but now they are only to be found on the Pacific Coast of the United States.

Redwoods grow so much because they use the fertile sediment that accumulates each year around their trunks due to the river swellings. The redwood can then form a new radicle system at the required level.

The redwoods also have a very special feature that enables them to survive the passing of time. They are very fire-resistant thanks to the spongy bark of their trunks, which can be up to 28 inches (70 centimeters) thick.

Redwoods might be the most voluminous trees, but the thickest tree in the world in the world is a cypress in Mexico, known as the Giant, with a circumference of 125 feet (38 meters) although it is "only" 154 feet (47 meters) high.

(1) A forest of giants
To walk through a redwood forest is an impressive feeling, surrounded by giants of over 400 feet (120 meters) and 1,000 tons.

(2) Jumping spiders
There are spiders that do not use cobwebs for hunting. They are jumping spiders that stalk their prey until they jump on them, just like tigers.

(3) A rucksack full of babies
Tarantulas live in all the continents. The females carry their eggs in a kind of knapsack and when the babies are born, they travel for a few days on their mothers' backs.

(4) Nettle diet
Here you can see a butterfly whose caterpillars only eat nettles. If you find nettles anywhere, whether it be beside the sea or 6,600 feet (2,000 meters) up, look carefully and you will find one of these butterflies.

(5) Passing years
Here you can see the growth rings of the trunk of a 100-year-old giant redwood.

5

LICHEN, MOSS, AND FERNS

In areas of moisture and in the forest shade there are very special plants that produce no flowers.

Lichen are a wide range of plants that are made up of fungi living in close association with a primitive green plant, normally algae. The fungi form the outer layer while the inner layer is made up of the cells of the algae surrounded by filaments of the fungi. This intimate process is called *symbiosis*. In lichen, the algae's photosynthesis give energy for the lichen to grow and reproduce; and the fungi give the algae the nutrients they need and protect them from climatic conditions.

There are thousands of different species of lichen. Some grow vertically and branch out, and others look like simple flat leaves. The way they hold on to the surface of the plant they live on also varies a great deal; for instance, there are lichen that form scabs on rocks or on tree bark.

Ferns and moss are plants that reproduce through small *spores*. They abound on forest floors, as they are usually damp and rich in mineral substances. These features tremendously aid their growth.

Have you ever noticed the strange "carpets" covering rocks or the base of tree trunks in the forest? These are moss—their stalks grow so close to one another that they form live rugs.

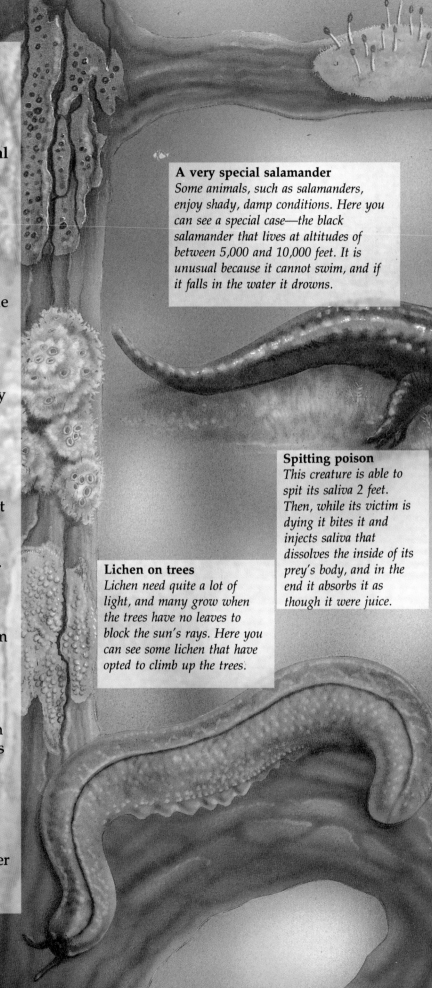

A very special salamander
Some animals, such as salamanders, enjoy shady, damp conditions. Here you can see a special case—the black salamander that lives at altitudes of between 5,000 and 10,000 feet. It is unusual because it cannot swim, and if it falls in the water it drowns.

Spitting poison
This creature is able to spit its saliva 2 feet. Then, while its victim is dying it bites it and injects saliva that dissolves the inside of its prey's body, and in the end it absorbs it as though it were juice.

Lichen on trees
Lichen need quite a lot of light, and many grow when the trees have no leaves to block the sun's rays. Here you can see some lichen that have opted to climb up the trees.

Small cushions
The Dicranaceae cirrata moss grows in small cushions. Its leaves are very twisted when it is dry but are spread when they absorb moisture.

"Vegetable scabs"
Some lichen form perfectly encrusted scabs on the surface of bark and on rocks.

Bags of spores
Ferns, such as this polypodium, reproduce through spores. Spores form in tiny bags called sporangia that group in packets on the lower faces of foliage.

Glossary

conifer a family of trees including evergreens and shrubs, some of which have pinecones or fruit

hibernate when some animals enter a state of sleep during the winter; by doing so they avoid the harsh weather conditions

humus brown or black material resulting from partial decomposition of plant or animal matter that forms the organic portion of soil

ichneumon a family of insects whose larvae are usually internal parasites of other insect larvae and especially of caterpillars

integument an enveloping layer of skin of an organism

invertebrates animals that do not have spinal columns, such as earthworms, arthropods, and mollusks

photosynthesis a process in which green plants synthesize organic material through carbon dioxide, using sunlight as energy

processionary caterpillar a caterpillar whose larvae make large webs on trees and move together in columns to feed

spores primitive single-celled reproductive bodies produced by plants that are capable of developing into new individuals, sometimes unlike the parent

strata a series of layers or levels in an ordered system

symbiosis the living together of two different organisms in a mutually beneficial relationship

taiga a moist subarctic coniferous forest that begins where the tundra ends and has numerous spruces and firs

undergrowth vegetation–generally seedlings, herbs, and shrubs–that grow on the ground of the forest

Index